D0846064

UNIQUE ANIMALS OF
THE SOUTH

By Tanya Lee Stone

BLACKBIRCH PRESS

An imprint of Thomson Gale, a part of The Thomson Corporation

THOMSON

™

GALE

Detroit • New York • San Francisco • San Diego • New Haven, Conn. • Waterville, Maine • London • Munich

Tusc. Co. Public Library
121 Fair Ave NW
New Phila., OH 44663

3 1342 00467 1656

© 2005 Thomson Gale, a part of The Thomson Corporation.

Thomson and Star Logo are trademarks and Gale and Blackbirch Press are registered trademarks used herein under license.

For more information, contact
Blackbirch Press
27500 Drake Rd.
Farmington Hills, MI 48331-3535
Or you can visit our Internet site at http://www.gale.com

ALL RIGHTS RESERVED.
No part of this work covered by the copyright hereon may be reproduced or used in any form or by any means—graphic, electronic, or mechanical, including photocopying, recording, taping, Web distribution, or information storage retrieval systems—without the written permission of the publisher.

Every effort has been made to trace the owners of copyrighted material.

Photo Credits: Cover: Corel; © W. Perry Conway/CORBIS; Corel, 5, 12, 13 (both), 14, 15 (below), 17, 22, 23; © Philip Gould/CORBIS, 18, 19; © Darrell Gulin/CORBIS, 9, 10; © Gavriel Jecan/CORBIS, 6; © George D. Lepp/CORBIS, 10; © Steve Maslowski / Visuals Unlimited, 21; © Joe McDonald/ CORBIS, 7; © Roy Morsch/CORBIS, 20; Photodisc, 15 (top); © Kevin Schafer/CORBIS, 3

LIBRARY OF CONGRESS CATALOGING-IN-PUBLICATION DATA

Stone, Tanya Lee.
  Unique Animals of the South / by Tanya Lee Stone.
    p. cm. — (Regional Wild America)
  Includes bibliographical references and index.
  ISBN 1-56711-968-9 (hardcover: alk. paper)
  1. Animals—Southern States—Juvenile literature. I. Title II. Series: Stone, Tanya Lee.
Regional wild America.
  QL157.S68S76 2005
  591.975—dc22
                                                                    2004018670

Printed in the United States of America
10 9 8 7 6 5 4 3 2 1

Contents

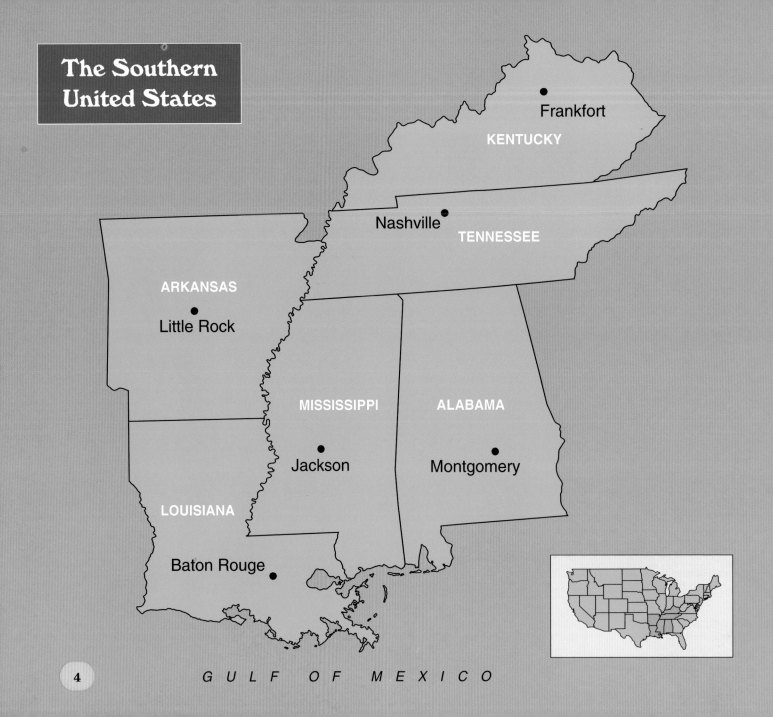

The Southern United States

Frankfort

KENTUCKY

Nashville

TENNESSEE

ARKANSAS

Little Rock

MISSISSIPPI

ALABAMA

Jackson

Montgomery

LOUISIANA

Baton Rouge

G U L F   O F   M E X I C O

In the South, birds fly, marine
life swims, and animals travel
across the land. Many different
animals make their homes here.
Some animals are especially well
known in this region.

A male white-tailed deer
stands by a river. The South is
home to a variety of animals.

Outgoing Otters

The river otter is a mammal that spends a lot of time in the water. These animals are found in rivers and streams. Otters prefer clean waters where there are plenty of fish to eat. They weigh between 10 and 30 pounds (4.5 and 14kg). An otter stretches 23 to 33 inches (58 and 84cm) in length. Its tail adds 10 to 20 inches (25 to 51cm).

River otters are playful and smart. Their bodies are designed for swimming.

A river otter's body is well suited for swimming. It is long and sleek. It uses its tail like a boat's rudder to steer. An otter's legs are strong and its feet are webbed. An inner coat of short fur keeps it warm. An outer coat of coarse guard hairs sheds water. An otter's eyes are positioned high on its head, which helps it see while swimming. While most of its body stays underwater, it can poke its head up out of the water to see. And like those of seals, an otter's nostrils close when it goes underwater.

Otters are smart, social, playful animals. Females teach their babies how to swim and hunt for fish. They do not build dens. Instead, they take cover in a natural shelter, such as a hollow log or opening in a ledge. They also take over other animals' old burrows or dens. An otter will occupy the den of a beaver or nutria.

An otter munches on a fish. Otters live in rivers and streams where they can find plenty of fish to eat.

Nutrias were brought to Louisiana from South America for their fur. Like otters, nutrias are skilled swimmers.

Nimble Nutria

Like otters, nutrias love the water. Nutrias are large rodents, related to beavers. They are native to South America and were brought to Louisiana to be used for the fur trade in the 1930s. Since then, they have spread to other parts of this region. Nutrias live in streams, ponds, and marshes. They are about 12 to 36 inches (30 to 91cm) in length. Their long tail adds another 12 to 14 inches (30 to 36cm). Nutrias weigh 10 to 12 pounds (4.5 to 5kg).

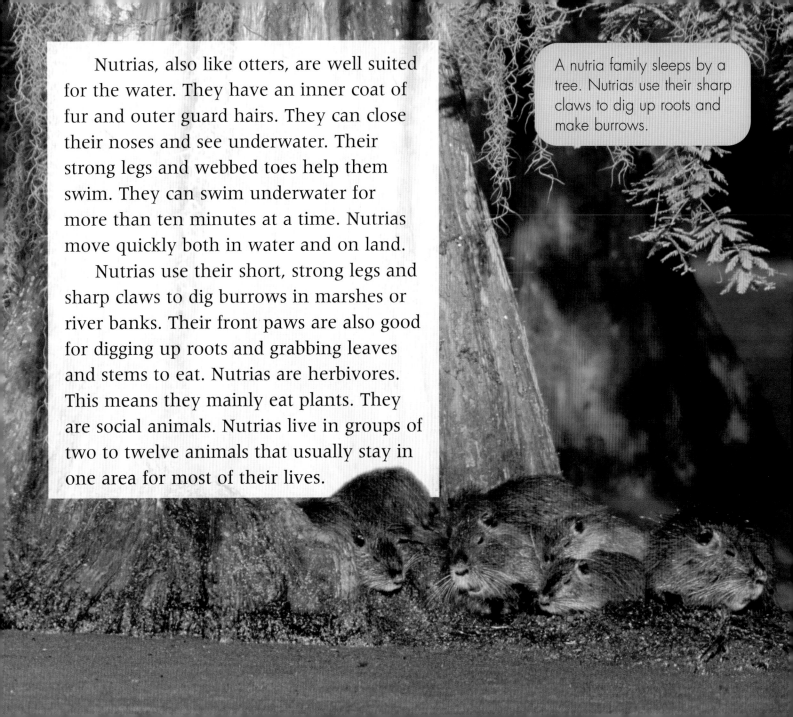

Nutrias, also like otters, are well suited for the water. They have an inner coat of fur and outer guard hairs. They can close their noses and see underwater. Their strong legs and webbed toes help them swim. They can swim underwater for more than ten minutes at a time. Nutrias move quickly both in water and on land.

Nutrias use their short, strong legs and sharp claws to dig burrows in marshes or river banks. Their front paws are also good for digging up roots and grabbing leaves and stems to eat. Nutrias are herbivores. This means they mainly eat plants. They are social animals. Nutrias live in groups of two to twelve animals that usually stay in one area for most of their lives.

A nutria family sleeps by a tree. Nutrias use their sharp claws to dig up roots and make burrows.

Simply Swamps!

Swamp rabbits are another water-loving animal. The swamp rabbit is a kind of cottontail, which means it has white on the underside of its tail. These rabbits prefer marshes and swamps, where there are plenty of plants to eat. They are herbivores. The swamp rabbit is the largest cottontail in North America. It weighs 4 to 6 pounds (1.8 to 2.7kg). Females and males are usually about the same size. Their bodies are 18 to 22 inches (46 to 56cm) in length.

A swamp rabbit rests near a marsh. These cottontail rabbits run zigzags and even swim to escape predators.

Like other rabbits, females give birth several times a year. Swamp rabbits can have up to five litters (groups of babies) each year. Each litter has between two and six babies. Female swamp rabbits build a nest with plants and fur. Unlike most rabbits, swamp rabbit babies are born with some fur and open their eyes within a few days.

These rabbits avoid predators (animals that hunt other animals for food) in much the same way other rabbits do. They either run away in a zigzag pattern that makes them hard to catch, or they stand perfectly still. These rabbits will also stay motionless in the water, with only their noses poking out to breathe! They are good swimmers, and will even dive underwater. Swamp rabbits are hunted by people, as well as many other animals, including alligators and bobcats.

Marsh rabbits like to eat swamp plants. Females also use the plants to build nests.

Bounding Bobcats

The bobcat is the most common wildcat in North America. But the largest populations are in the South and Southeast. Bobcats usually weigh between 10 and 35 pounds (4.5 and 16kg). They are about 24 to 48 inches (61 to 122cm) long. Unlike most cats, bobcats have fairly short tails. This cat's 4 to 6 inch (10 to 15cm) bobbed tail gives the animal its name.

Bobcats are solitary animals. They generally live alone, except during mating season. They are usually nocturnal, which means they are mainly active at night. Bobcats are excellent hunters. These meat eaters hunt for many kinds of prey (animals hunted by other animals for food). They eat rabbits, mice, squirrels, rats, sheep, goats, and other animals. They will also hunt reptiles, amphibians, fish, and birds. Like other cats, they stalk their prey and then they pounce! Bobcats do not have many predators. But wolves and coyotes will prey upon them and their babies. Foxes and owls will also kill bobcat kittens.

Female bobcats can give birth once a year. They usually have one to three kittens. A female raises her kittens without the male. Kittens are born with their eyes closed. Their eyes open several days later. Their mother nurses the kittens for seven or eight weeks. Young bobcats are ready to go off on their own within a year.

Bobcats like to hunt alone at night. Baby bobcats grow up and leave their mother within a year.

White-tailed deer are common in the meadows and woodlands of the South.

Delightful Deer

The white-tailed deer is well named. It has a brown tail with white edges and a white underside. White-tailed deer flash the underside of their tail to send a danger signal. This is called tail-flagging. Female deer also flag their tails to help their babies (fawns) follow them through the forest. A fawn's coat is spotted with white. This helps protect the fawn, because predators often think the markings are just spots of sunlight in the forest.

The white-tailed deer lives in many parts of the United States. It is a very common sight, though, in the South. In fact, the white-tailed deer is the state mammal of both Mississippi and Arkansas. Deer live in meadows and woodlands. They are herbivores. They eat shoots, buds, leaves, and stems. They nibble on nuts and fruit. In winter, deer eat bark and tall brush. They also dig under the snow to uncover any food they can find.

Male white-tailed deer often live in small groups. The white markings of a fawn's coat (below left) help it hide from predators in the forest.

Deer are shy, gentle animals. Female deer are called does. Male deer are called bucks. Does and fawns often live and feed together in small groups. Bucks also form small groups. White-tailed deer have excellent senses of sight, smell, and hearing. They use these senses to know when danger is near. Deer are swift runners and move quickly to escape predators.

Pouched 'Possums

The opossum is a marsupial. Other examples of marsupials are kangaroos and koalas. The opossum is the only marsupial that lives in North America. Marsupials are unique mammals. A female marsupial has a pouch on the outside of her body. Baby opossums (called joeys) are tiny, deaf, blind, and furless. As soon as they are born, they must climb into their mother's pouch. There, they stay warm and fed. Joeys nurse for two to four months in the safety of the pouch. Once they leave the pouch, joeys travel on their mother's back for another two months. Then they are ready to go off on their own.

Adult opossums weigh between 4 and 14 pounds (2 and 6kg). They are 13 to 20 inches (33 to 51cm) long. Their long tails add another 10 to 21 inches (25 to 53cm). Opossums have thick undercoats to keep them warm and coarse topcoats to help keep them dry. They are good swimmers and good climbers. They build dens or nests both on the ground and in trees. Opossums will also take over the dens of other animals.

Opossums eat many kinds of foods and are well suited for hunting. They have sharp claws and special tails that help them balance in trees. Their keen senses of smell, hearing, and sight help them hunt at night. They have many sharp teeth, and mouths that can open very wide to fit a variety of foods. Opossums are omnivores (eating both plants and animals) and eat almost anything they can find. They like fruit, nuts, and insects. They also hunt frogs, mice, rats, and crayfish. To ward off predators, an opossum will play dead, a habit that led to the expression "playing 'possum." An opossum may also bare its teeth, hiss, or drool to look sick. Many animals are not interested in eating sick or dead animals. When they see an opossum behaving this way, they leave it alone.

Baby opossums (left) stay on their mother's back for about two months before going out on their own. Adult opossums (above) have sharp teeth and a wide mouth for eating a variety of foods.

Crawling Crayfish!

Like crabs and lobsters, crayfish are crustaceans. A crustacean is an aquatic animal with an exoskeleton (its shell) and a jointed body. The shell helps protect its body. Unlike their relatives, crayfish live mainly in freshwater. They live in rivers, streams, ponds, and marshes. They are very common in the South. In fact, the crayfish is the state crustacean of Louisiana. Crayfish are also called crawfish and crawdads. A favorite treat for humans, crawfish can be found on menus all over the South. Many animals prey upon them, too. Otters, herons, fish, and snakes all eat crayfish.

Crayfish are freshwater crustaceans often found on southern menus. Below, a female curls her tail around her cluster of eggs.

Crayfish use their antennae to sense their surroundings and find food. Crayfish eat many kinds of food.

Crayfish are omnivores and eat many kinds of water plants and animals. They are nocturnal, and mainly hunt at night. Crayfish like to feed on worms, snails, shrimp, and small fish. Their two pairs of antennae (one long and one short) help them sense their surroundings and find food. Like lobsters, crayfish have eight jointed walking legs. Their legs are also used to reach into small crevices and locate food. If a crayfish loses one of its legs, it can grow another to replace it. The two large claws of a crayfish are similar to those of a lobster. Crayfish are much smaller than lobsters, though. They are only about 2 to 6 inches (5 to 15cm) long.

Beautiful Bass

Largemouth bass swim in lakes, ponds, rivers, marshes, and swamps. They prefer warm, shallow waters. In the South, the largemouth bass is the state fish of Arkansas and Mississippi, the state sport fish of Tennessee, and the state freshwater fish of Alabama. This fish has a long, torpedo-shaped body. Adults are 9 to 25 inches (23 to 64cm) in length. They can weigh more than 20 pounds (9kg). Its large mouth gives this fish its common name. A largemouth bass's jaw often stretches past its eye.

A largemouth bass fights, hooked on a fishing lure. This fish gets its name from its large mouth.

This fish is often an olive green color on its top side. A dark marking lines its side from head to tail. On its underside, a largemouth bass is light green to white. The dorsal (top) fin has a deep notch that almost breaks it into two parts—the front with spines and the back with softer parts.

Young largemouth bass eat insects, plants, and small fish. Adults can weigh more than 20 pounds.

Young largemouth bass feed on insects and small fish. They also eat tiny plants. Adults eat other fish, frogs, crayfish, and insects. The largemouth bass's main predators are herons, kingfishers, and humans. In America, the largemouth bass is the most popular species fished for sport. These fish are plentiful. One reason for this is that females lay thousands and thousands of eggs. The eggs are then guarded by males. After they hatch, the male protects the mass of tiny fish. He scares off predators until the fish are ready to swim off on their own.

A big gobbler struts in a snowy forest. Male turkeys have colorful feathers.

Turkey Tails

Wild turkeys are related to pheasants and grouse. They are found in many parts of the eastern half of the United States. In this region, they are plentiful. They are also the state game bird of Alabama. Males are called gobblers or toms. Females are called hens. Adults grow to be about 4 feet (1.2m) tall. Gobblers can weigh more than 20 pounds (9kg), while hens do not usually weigh more than 12 pounds (5kg).

Male turkey feathers are more colorful than a female's. A hen's feathers are brown or gray. But a gobbler's feathers are copper, red, bronze, black, green, and gold! Turkeys have long tail feathers that stretch 12 to 15 inches (30 to 38cm). Toms can spread their tail feathers like a fan. They often do this to attract the attention of a hen. Turkeys have reddish legs with clawed toes. Males grow spurs on the backs of their legs. These sharp points are used for fighting.

Wild turkeys have keen hearing and vision. They can move quite quickly to escape danger. These turkeys can run more than 20 miles (32km) per hour and fly more than 50 miles (80km) per hour! Wild turkeys are hunted by people. But they are not the same kinds of turkeys that are raised to end up on our dinner tables.

There are many unique and wonderful animals that live in the South. All of them add to the richness and beauty of this region.

A tom turkey fans out his tail feathers to attract the attention of a hen.

Glossary

Herbivore An animal that mainly eats plants.
Omnivore An animal that eats plants and other animals.

Predator An animal that hunts another animal for food.
Prey An animal that is hunted by another animal.

For More Information

Lee Jacobs, *Opossum*. San Diego, CA: Blackbirch Press, 2003.

Tanya Lee Stone, *Living in a World of Brown*. San Diego, CA: Blackbirch Press, 2001.

Stephen Swinburne, *Bobcat: North America's Cat*. Honesdale, PA: Boyds Mills Press, 2001.

Sally Tagholm, *The Otter*. New York: Kingfisher, 1999.

Index